Secrets of Nature
FROM C-Z

DR. TUGLER

Copyright © 2024 by Dr. Tugler. 857770

All rights reserved. No part of this book may be reproduced or transmitted in any form or by any means, electronic or mechanical, including photocopying, recording, or by any information storage and retrieval system, without permission in writing from the copyright owner.

To order additional copies of this book, contact:
Xlibris
844-714-8691
www.Xlibris.com
Orders@Xlibris.com

ISBN: Softcover 979-8-3694-2704-0
      EBook   979-8-3694-2705-7

Library of Congress Control Number: 2024914514

Print information available on the last page

Rev. date: 08/07/2024

Contents

Closer is Colder, Farther is Hotter 1

Construction – Serious Business 5

Catastrophe in Space 9

Cars 14

Capillaries 18

Conversation 22

Let's continue with letter
C

Other books by the author

Secrets of Nature from A-Z
Secrets of Nature from B-Z
Secrets of Nature from C-Z

Coming soon: *Secrets of Nature from D-Z*

Dear parents,

You could read this story with your smart, curious children who are interested in investigation, science, and nature. Teachers could use it in class to read and analyze different situations with the students. This story is for young adults and others, especially those interested in science, solving problems, and investigation.

To all readers: This book could help improve your knowledge of different realms of nature and sharpen your intellectual skills.

In the books you will get acquainted with space pirates and students of Galactic University, smart octopi and even smarter whales, disputing chemists, and oil-miners! These and other book characters will take you to space and underground, to the North Pole and to the African desert, to mountains and waterfalls, to warm snow and cold fire, and even to listen to an echo. All these fantastic adventures are waiting for you in the series.

Closer is Colder, Farther is Hotter

We all know that the sun is the source of life on the earth. But not everybody knows how the earth actually interacts with the sun or what processes are involved in this interaction.

Robert, lying on warm grass on a sunny spring day, started to think about why it was cold in the winter and why it was hot in the summer. His teacher at school had been explaining something about the sun and the earth, but no matter how hard Robert tried to understand it, he just couldn't. After considering everything he could remember, he decided to ask his father. "He will definitely make the whole thing clear. I am pretty sure that it's impossible to turn the sun's heat on and off as we do in our house," Robert thought.

Robert: If I sit near a fire, where will it be hotter: closer to the fire or farther away from it?

Dad: When you sit closer to it. A greater flow of energy falls on you.

Robert: What energy are you talking about?

Dad: I mean the energy of heat. Infrared rays spread out as they radiate.

Robert: What are infrared rays?

Dad: Infrared rays are the rays of heat. They are invisible.

Robert: Alright. Now, is the earth closer to the sun in the winter than in the summer?

Dad: The earth is closer in the winter than in the summer.

Robert: What? But it's much colder in the winter than in the summer. You are contradicting yourself.

Dad: What do you mean?

Robert: Closer to the sun is colder, farther away is hotter? How do you explain this?

Dad: You see, the heating of any surface caused by a point source of light depends not only on the distance between them, which you are experiencing right now, but also on the angles of incident rays - I mean the angles between those rays which strike the surface and the lines perpendicular to the surface itself.

Robert: And what is a point source of light?

Dad: It's a source of light with dimensions that are of no significance because you are far away from it. I mean you are so far away that the dimensions of the light source are negligible.

Robert: Are you kidding? How could anyone consider *the sun* to be a *point* source of light?

Dad: We can consider the sun to be a point source of light because its dimensions may be disregarded in comparison to the distance between the sun and the earth. We're a long way from the sun, really very far away from it - so far that its size actually makes no difference. It radiates as if it were only a tiny dot.

Robert: So, why do you say that the amount of heat the earth's surface receives is different in winter and in summer?

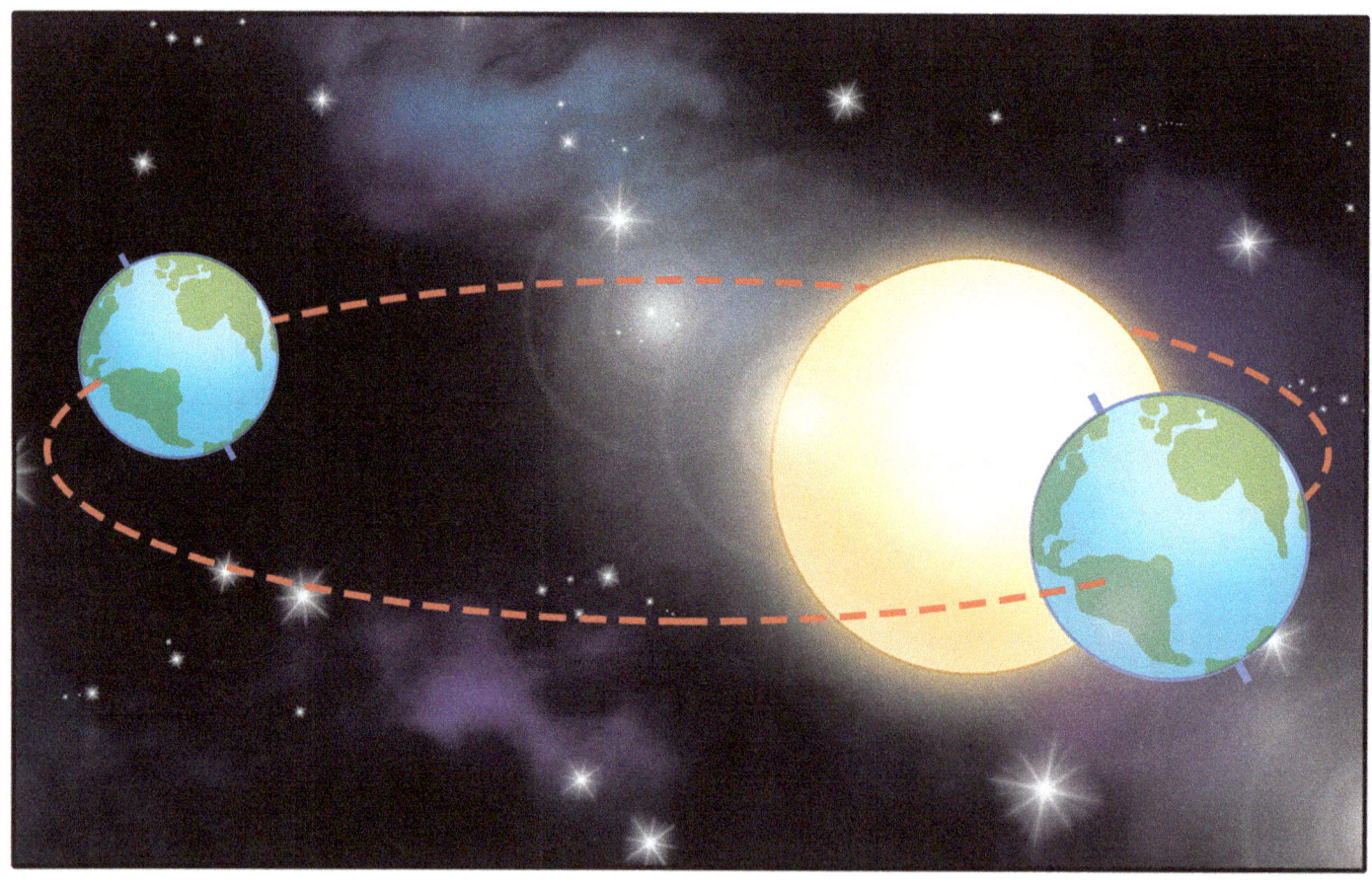

Dad: The earth has an axis of rotation which is tilted to the plane of the earth's orbit. Though in the winter the earth is nearer to the sun than in the summer, the angle of incidence in the Northern Hemisphere of the earth, where we live, is greater than in the summer. The rays skim over the surface, and it means that the earth receives less energy in the winter than in the summer. It's the tilt that counts.

Robert: You're saying that this tilt is more important than the difference in the distance between the earth and the sun. Right?

Dad: Absolutely right! There're no more contradictions, right? Now, do you understand why sometimes nearer might be colder, and farther away might be hotter?

Robert: I think I do. It's much easier than I thought in the beginning.

Dad: You are right. The world is built in an ingeniously simple way.

Robert: Thanks. Now I understand you. I just wonder if you have persuaded our readers as well.

Dad: Let's ask them.

Robert: Okay.

Dad: Dear readers, do you agree with our solution, or do you have your own theory about why it is colder in the winter and hotter in the summer? And what do you think is going on in the Southern Hemisphere?

Robert is smiling. Is it possible to find anybody, even in the Southern Hemisphere, who would not agree with his dad?

Dear parents,

For your children to understand this story clearly, we prepared investigative questions and answers for them. You can use the answers if you need them to discuss "Closer is Colder, Farther is Hotter" with your children. But first ask them to try to find the answers in the story.

Dear Children! Are you a detective? Find the answers to the questions in the text.

Questions

1. When you sit near a fire, where will it be hotter: close to the fire or farther away from it?
2. What are infrared rays?
3. When is the earth closer to the sun: in the winter or in the summer?
4. Why is the sun considered to be a point source of light?
5. Why is the amount of heat the earth's surface receives different in winter and in summer?

Answers

1. Close to the fire.
2. Infrared rays are rays of heat.
3. In the winter.
4. The sun's dimensions may be disregarded in comparison to the distance between the sun and the earth. We're so far away from the sun that its size actually makes no difference. It radiates as if it were only a tiny dot.
5. Because of the angle of the light going to the earth from the Sun and the tilt of the axis.

Construction – Serious Business

In the depths of space, the dwellers of the planet Gamma from the star Altair used different mechanisms such as pulleys, levers, screws, inclined planes, and more to assist in their construction.

One day, everything was normal, familiar, even a bit boring. Suddenly, the construction workers saw that between two heavy metal gates that were closing, a mail ship – a small spaceship that travels short distances and delivers mail – was stuck. The ship's crew consisted of a single carrier, a man from planet Earth, who delivered very important packages: presents for birthdays and items from family and other loved ones whom the construction workers rarely saw. Two of the construction workers, Green and Mahogany, rushed to the metal gates, which continued to threaten the mail ship. A little more, and the gates would crush the ship like paper! Yes, in space you must always be on the lookout for danger! They needed to quickly save the mailman because there was no time to get help. It was necessary to move the gates to the side without special tools, with only the help of pulleys and ropes that were attached to the belts of Green and Mahogany. They needed to quickly solve the problem: what is the force with which the construction workers pull the rope, transferred through a movable pulley and tied to Green's belt, if the weight of the humanoid is seven hundred newtons? Think fast, construction workers, because the fate of the mailman rests in your hands!

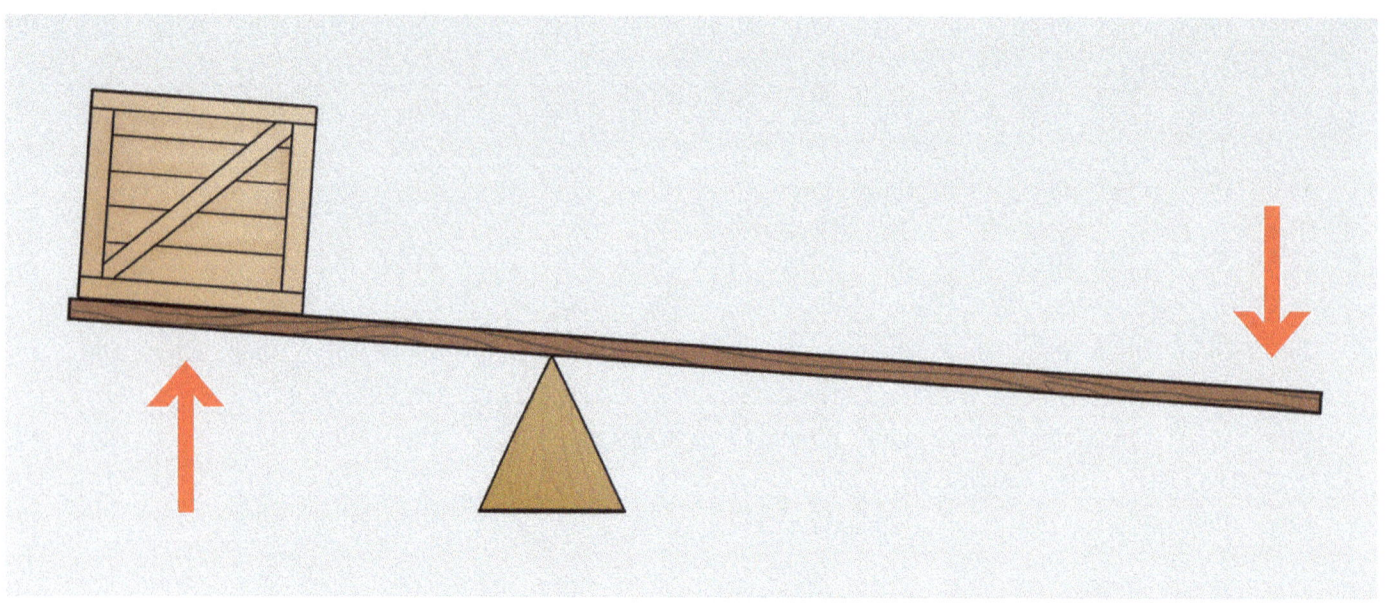

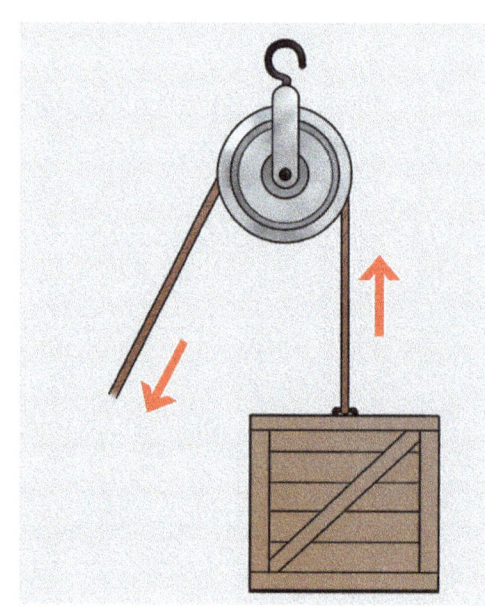
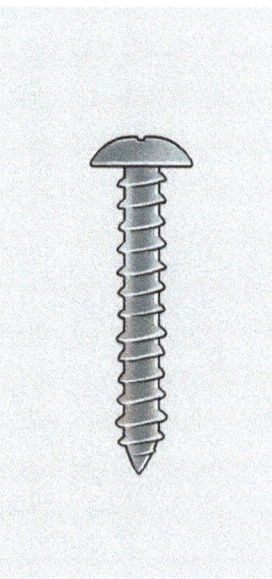

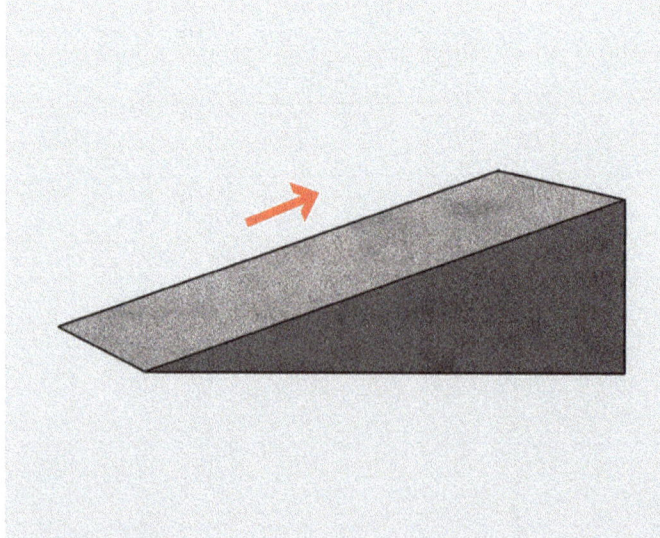

Green: I think that to solve this important problem, you should do an experiment and see what happens.

Mahogany: Experimenting on living humanoids is illegal.

Green: Then we need to find volunteers.

Mahogany: Instead, let's do an experiment in a laboratory, where we create a similar environment to the situation at hand.

Green: We don't have time to organize a laboratory, especially when we are in a hurry to save the mailman's life! Let's try here instead. We can take an immovable pulley and throw the rope through it. We attach the rope to a movable pulley and link the load of seven hundred newtons to it. We attach a dynamometer onto the immovable pulley.

Mahogany: What will the dynamometer show?

Green: Think about it theoretically. The effect of the force of gravity is equalized by the sum of the forces of elasticity from the ends of the rope, therefore the force of gravity of each end will be half the weight of the load.

Mahogany: So the dynamometer should show three hundred and fifty newtons.

Green: Let's check! Yes, it is!

Mahogany: Do you think, if we hang another dynamometer, it will show the same value?

Green: Let's test it.

Mahogany: Of course we can check but I would rather hurry up our experiments; we need to know, in advance, what will happen in any unexpected case.

Green: I think that the dynamometers will show the same results.

Mahogany: It would be good to get advice from the Earth inhabitant since they are great at physics.

Green: There's nothing to get advice about, both ends of the rope can be pulled identically.

Mahogany: We still need to prove it.

Green: Alright, let's start!

Mahogany: Great! We figured it out, just like the Earth humanoids!

Well, dear readers, let's not disgrace the honor of Earth inhabitants! We can, with you, predict the result in advance, without experiments, which would make Green and Mahogany extremely joyous!

Dear parents,

For your children to understand this story clearly, we prepared investigative questions and answers for them. You can use the answers if you need them to discuss "Construction – Serious Business" with your children. But first ask them to try to find the answers in the story.

Dear Children! Are you a detective? Find the **a**nswers to the questions in the text.

Question

What are pulleys, levers, screws, and inclined planes?

Answer

They are simple machines.

Catastrophe in Space

In a huge room with a round table, experts in the space industry gathered to discuss space problems. Everyone in attendance was silent and listening hard, trying to determine voices in a recording. This recording was the only thing left of a destroyed spaceship. They were analyzing the errors and mistakes that caused the catastrophe due to a bad understanding of the laws of nature. These mistakes must never be repeated…

Recording: We cannot identify what is happening; are we moving or not? Our measuring instruments aren't working! Strange things are happening in the ship! HELP! SOS! SOS!

President: (Addressing the experts) You just listened to the recording of the events on the ship. Now I would like your opinion on the incident.

Expert: The astronauts couldn't tell if they were moving or not. The problem is clear: they didn't have any reference bodies.

President: What do you mean? What bodies were they supposed to have? Did they leave something on Earth when they took off? Were they poorly equipped? Are the companies that equipped them to blame? Should we sue them?

Expert: Every movement is relative, and if there are no bodies that are relative to the motion, then it's impossible to figure out if the object is moving or not. For example, Earth moves in space at a high speed, yet on Earth no one notices the movement. Do you agree?

President: They should have listened to the sound of the engine.

Expert: How would they hear the engine when all around them was airless space and sound moves only in an environment, for example, in the air?

President: Wait, that's right. When we were on the Moon, we did not hear anything without special equipment. Information only traveled through wires. Sound did not spread in the vacuum. But let's continue listening. Here's the recording of the voice of the commander of the ship: "The motion graph appeared on the display: a graph of velocity relative to time."

Expert: Mr. President, do you have an image of the graph?

President: Here is the image, please put it on the display.

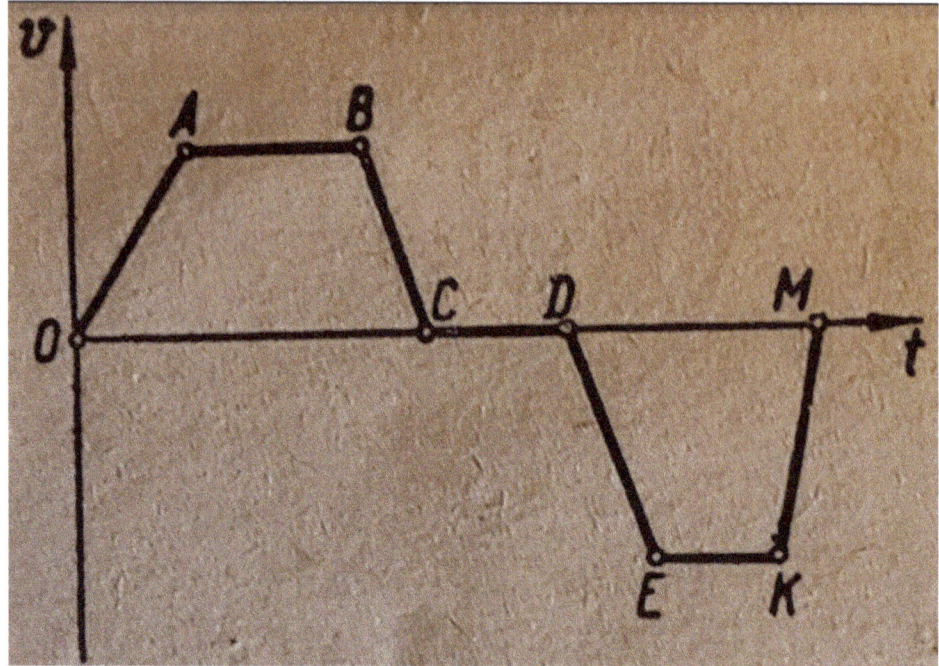

Expert: Yes, that's complicated motion! I think that from the station, the ship first accelerated (DA) and then flew at a constant speed (AB).

President: It seems you're correct. Then the ship slowed down equally (BC) until it reached a stop and stood still for some time (CD).

Expert: Oh, no! The ship started accelerating and moving backwards in the opposite direction (DE). Then the ship moved backwards at a constant speed (EK). It seems to me that there was something dangerous ahead.

President: Yes! Yes! It kept moving backwards, but decelerating until stopping for a split second before slowly accelerating forwards (KM).

Expert: I know that only the commander of the spaceship could make quick and correct decisions. During high pressure moments, his brain worked amazingly well and very precisely, like a supercomputer. It's a pity to lose such a good commander…

President: Let's finish analyzing the spaceship recording.

Expert: Look, there's another graph. It's the displacement of the spaceship in space.

President: Yes…they made some risky maneuvers. It seems like the built-in computer was not able to navigate the ship properly and the commander decided to pilot the spaceship manually.

Expert: Mister President! We must equip our spaceships better! The astronauts require more equipment for survival!

President: What else do they need to have?

Expert: A ball. Yes! Just a ball!

President: We have no time for your jokes right now!

Expert: Let me explain. If you put a ball on a table…although, wait. They need a special machine that creates artificial gravity in the spaceship.

President: They had such a machine.

Expert: They can use the ball as a tool to understand how and in which direction the ship is moving. If the ball stays on the table without moving, that means that the spaceship is moving at a constant velocity. However, if the ball slides forwards in the direction that the ship is moving, then that means that the ship is slowing down. If the ball rolls back, in the opposite direction of the ships movement, then the ship is accelerating.

President: What if the ball rolls to the side? How is the ship moving then?

Expert: I think that means the spaceship is turning to the right or to the left. This would have given the commander an idea of how the spaceship was moving! We must immediately equip every spaceship with a ball!

President: Yes, we now know that they died from too much force pushing down on them, and their hearts could not handle it.

Expert: We can make a virtual model that will show us the movement of the ball on our computers.

President: More computers?! We've already seen that they are unreliable tools that break often.

Expert: I think that the ball and a computer can easily replace tools such as the level. A level is a tool with a liquid inside and an air bubble that moves around in the liquid. The bubble moves similarly to a ball in a spaceship. The only downside to a level is that it does not show when the spaceship is turning.

President: Therefore, the ball is a much more universal and useful tool.

Expert: That is what many experts have decided as well.

President: Agreed. We'll settle with the ball idea. Gentlemen, it is my pleasure, as leader of this company, to thank you all for your help in our discussion of the problem regarding the safety of spaceships. Hopefully, with your help, we will be able to stop any future catastrophes. If you have any other ideas regarding the topic, please let me know. We must do everything in our power to keep our spaceships and astronauts safe.

The President and the group of Experts took a moment of silence to remember the brave men who died on the spaceship, hoping the event would never repeat again. As they stood, the recording stopped, and the only thing that could be heard was the sound of the machine turning off.

Dear parents,

For your children to understand this story clearly, we prepared investigative questions and answers for them. You can use the answers if you need them to discuss "Astronauts" with your children. But first ask them to try to find the answers in the story.

Dear Children! Are you a detective? Find the answers to the questions in the text.

Questions

1. What happened to the astronauts who were flying in space?
2. Why were the astronauts unable to tell if they were moving?
3. Why couldn't the astronauts hear the engine?
4. What is shown on the graph that the experts have?
5. How does the spaceship move?
6. Why did the experts and the president decide to equip spaceships better?
7. What kind of equipment do astronauts require for survival in space?

Answers

1. Their spaceship was destroyed, and the astronauts died.
2. They did not have any reference bodies.
3. Because there is no air in the spaceship for the sound to travel through, and sound needs a medium to travel through.
4. They have a graph of the speed of the spaceship changing as a function of time.
5. The ship first accelerated, then flew at a constant speed, then decelerated, stopped, started accelerating in the opposite direction, flew at a constant speed, and decelerated in the original direction.
6. The astronauts require more equipment for their survival!
7. A simple ball, a level, or a virtual model on a computer.

Cars

Testers are people who test new models of machines and mechanisms. The reliability of the different kinds of machines is very important. Some people test airplanes. These people are called test pilots. There are also racers who test a car's speed and endurance in competitions. For racers, the safety issues of cars are of utmost importance, especially when it comes to collisions. The **Pilot** and **Racer** decide to argue about which cars are reliable. Of course, the **Racer** knows more about cars, but the **Pilot** can see everything from above.

Racer: Let's make a bet!

Pilot: Fine!

Racer: What are we betting?

Pilot: Our new machines.

Racer: Your plane versus my car?

Pilot: Agreed.

Racer: Let's decide who drives which car.

Pilot: I'll take a sedan since I'll win anyway. Dexterity and courage are the most important traits of a tester's character, not the type of car.

Racer: It seems like you don't understand what we are arguing about.

Pilot: You think that a truck is stronger?

Racer: Of course it is!

Pilot: I disagree. According to Newton's 3rd law, the forces on the two cars during a collision should be equal, therefore they should be damaged equally.

Racer: Correct, but you're forgetting about the masses of the two cars. Since the sedan has a smaller mass, it will accelerate more during the collision and, therefore, be more damaged. Also, different types of cars are used for different goals.

Pilot: Agreed. When transporting a smaller amount of goods over a short distance, it is, in fact, most efficient to use a sedan. However, trucks are better when you need to transport heavier goods. They require fewer drivers and consume less fuel per each unit of cargo.

Racer: Is there a limit to the load that an automobile can carry?

Pilot: Overweight automobiles are limited by the roads that they drive on.

Racer: The greater the mass of the automobile and the faster it's moving, the more damage it does to the roads.

Pilot: However, roads need to last for a long time.

Racer: Therefore, the pressure of the tires on the road (mass of the automobile) should not be greater than allowed. Usually, it should 60-70 newtons per centimeter squared.

Pilot: But that doesn't mean that you can't increase the mass of the automobile and cargo?

Racer: Correct. You can increase the mass without changing the pressure on the road.

Pilot: And how do you do that?

Racer: First, you can reduce the weight of the automobile. To do this, the material that the car is made of should be changed for a lighter and more sturdy material.

Pilot: Understood. So the decrease in the weight of the automobile allows for more cargo.

Racer: Second…can you guess? You are standing on the floor, right?

Pilot: You see that yourself.

Racer: Without moving, try to double your pressure on the floor.

Pilot: (tries it) Easy! Just stand on one leg! Here's another solution: come here and I'll pick you up. Oof! You're heavy!

Racer: See, you already understand the second option: increase the surface area between the automobile and the road.

Pilot: Yes, you can make an automobile have 3, 4, or even 5 axles. You can also add an extra trailer with multiple axles or use wider tires.

Racer: Do you know how cars drive across marshes or loose snow? On pneumatic rollers! Pneumatic rollers are vehicles with large rubber barrels for wheels.

Pilot: The scariest automobiles are trucks.

Racer: Racing cars are scarier than trucks.

Pilot: Perhaps we should conduct an experiment on the models of the cars.

Racer: Scared?

Pilot: Who, me?!

Racer: I'll crush your car like glass!

Pilot: We'll see…What time are we meeting for the experiment?

Racer: Tomorrow at 10 in the morning.

Pilot: Okay, I'll be there.

Racer: Do we need referees?

Pilot: Why? We aren't dueling.

Racer: Fine. Be ready! See you tomorrow.

Pilot: See you tomorrow.

What will the experiment show? Who will win the bet? On whom does victory depend: the person behind the wheel or the car itself? We'll see tomorrow…

Dear parents,

For your children to understand this story clearly, we prepared investigative questions and answers for them. You could use the answers if you need them to discuss "Cars" with your children. But first ask them to try to find the answers in the story.

Dear Children! Are you a detective? Find the answers to the questions in the text.

Questions

1. What is a tester?
2. Who are Pilot and Racer?
3. Why is safety and reliability so important for cars?
4. What are the Racer and the Pilot betting about?
5. What is Newton's 3rd law in physics?
6. What is the difference between sedans and trucks?
7. What are the benefits of a truck?
8. What should you do to reduce damage done to roads?
9. How do you decrease the pressure by the cars on the road?
10. How does speed affect the pressure of the car?
11. What experiment do the Racer and the Pilot plan on conducting?

Answers

1. A tester is someone who tests new models of machines and mechanisms.
2. Pilot is a person who flies and tests planes, and Racer is a person who drives and tests different new cars.
3. Cars need to be reliable and stable to avoid collisions and protect drivers and cargo while also not destroying roads.
4. The Racer and Pilot are betting about which car can better survive a collision and which cars are better for using on roads and transporting cargo.
5. Newton's 3rd law states that the forces on both cars should be equal in a collision. However, the acceleration is indirectly proportional to mass; therefore, the more the mass of the car, the less damage.
6. They have different masses.
7. Trucks can carry more cargo, whereas you would need more sedans and drivers to carry the same amount of cargo.
8. Decrease the pressure on roads.
9. Make cars out of lighter materials and increase the number of axles and wheels on a car to increase the area of contact between the car and the road.
10. Since roads are not completely smooth, when a car is moving, pressure on the road is constantly changing, which destroys roads. When the speed of the car increases, the pressure changes faster, destroying roads even more.
11. They plan on crashing a truck and sedan into each other to see what happens in the collision.

Capillaries

In the hospital, two doctors are having a discussion. The topic of conversation for the average listener sounds a little strange: capillaries. However, the doctors are extremely excited, exclaiming and waving their hands. Perhaps this topic is very important to them?

Dr. Doodle: Today I was in the hematologist laboratory when I was once again surprised at how well the capillary tubes work there!

Dr. Salvador: What would we do without them?

Dr. Doodle: It's so interesting. How high can liquids rise in capillaries?

Dr. Salvador: I think that depends on a few things: what type of liquid is in the capillaries, what its density is, what the diameter of the capillary is, and what the acceleration of gravity is on the given planet.

Dr. Doodle: By the way, do you remember that we are preparing to travel to another planet?

Dr. Salvador: Of course, I never forget anything.

Dr. Doodle: So, since the inhabitants of the other planet have different blood compositions from ours, will we be unable to use our capillaries?

Dr. Salvador: The only thing I definitely know is that on a different planet, the height of the liquid in the capillaries will change. Since the height of the liquid in capillaries depends on multiple factors, we can calculate the height using a special formula.

Dr. Doodle: What is this formula?

Dr. Salvador: It is $h = 2\sigma/\rho g r$, where h is the height of the liquid in a capillary, σ is the coefficient of surface tension of a liquid, ρ is the density of the liquid, g is the acceleration of gravity on the planet, and r is the radius of the capillary.

Dr. Doodle: Based on the formula, we must keep in mind that the acceleration of gravity on the other planet is different.

Dr. Salvador: Also, did you know that if you take a capillary and put one end in the liquid while bending the other side, then the liquid will stay inside?

Dr. Doodle: One time, when I put a piece of fabric on a wet windowsill, it started dripping water from the bottom of the fabric. Why is that? Aren't there capillaries in fabric as well?

Dr. Salvador: Even though the fabric is bent, the water still drips down because of the force of gravity.

Dr. Doodle: Interesting. By the way, do you remember that the word capillary translates to a "strand of hair" in Latin?

Dr. Salvador: Yes; capillaries are so thin, just like hair.

Dr. Doodle: Did you also know that many projects about perpetual motion machines are based off the principles of the functions of a capillary?

Dr. Salvador: I have heard of one of them: the water-capillary perpetual motion machine proposed by Sinclair in the eighteenth century. The inventor assumed that the water from upper vessel A moving through siphon C into lower vessel B will return to the upper vessel through a tube D, which is a so-called capillary having a very small diameter. However, the expected movement of the water through the capillary did not occur. Do you know what was wrong?

Dr. Doodle: Based on the laws of capillaries and engines, the machine should work! What was the issue?

Dr. Salvador: The issue was that in the water-capillary perpetual motion machine, water rises to the upper vessel provided that the vessel is empty. However, the vessel wasn't empty, and the capillary tube expanded, so the water stopped and the machine didn't work.

Dr. Doodle: So the system invented by Sinclair doesn't work?

Dr. Salvador: Exactly. If there's even a little bit of stored liquid in the upper vessel, then the capillary tube ends up being just another way for the water to flow downwards, and the system will definitely not work.

Dr. Doodle: That is a pity. The idea was brilliant.

Dr. Salvador: I know a lot about capillaries, except the main idea: why does liquid rise in capillaries? I know that in the human body, there are about 10 billion tiny blood capillaries. When combined, the overall length of these capillaries reaches around 80 to 100 thousand kilometers. There are enough capillaries in our bodies to wrap around the equator twice if the capillaries were stretched out in a line.

Dr. Doodle: Your knowledge of capillaries in the human body rivals that of an encyclopedia.

Dr. Salvador: You exaggerate. Here's another fact: the capillary phenomenon is related to the hydrophobic and hydrophilic phenomena of liquids. During the process of hydrophilia, liquid within a capillary rises because of the force of surface tension on the liquid.

Dr. Doodle: Of course! And, logically, liquid in a capillary should stay lower than the liquid level in the vessel instead of rising during a hydrophobic process.

Dr. Salvador: I wonder, what other capillary phenomena exist in nature and everyday life? After all, the two of us are used to encountering them only in the medical field…

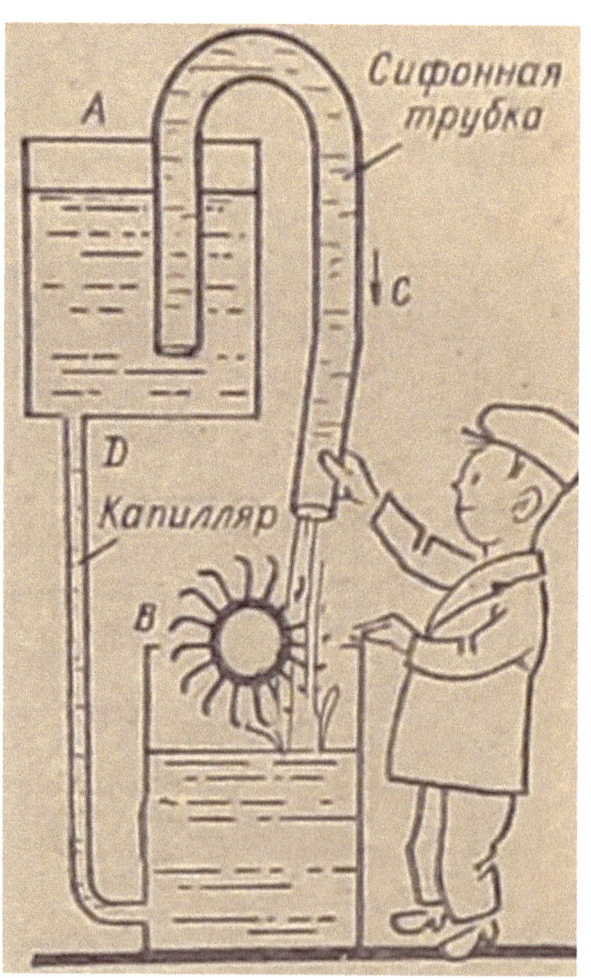

Dear parents,

For your children to understand this story clearly, we prepared investigative questions and answers for them. You could use the answers if you need them to discuss "Capillaries" with your children. But first ask them to try to find the answers in the story.

Dear Children! Are you a detective? Find the **a**nswers to the questions in the text.

Questions

1. What is a hematologist laboratory?
2. What are capillaries?
3. What factors affect how high a liquid can rise in a capillary?
4. How does the height of a liquid in a capillary change depending on the planet you are on?
5. Why does water drip from a bent cloth on a windowsill?
6. What is a perpetual motion machine?
7. Why does the system invented by Sinclair not work?
8. Why does liquid rise in capillaries?
9. How many micro-capillaries are in the human body?
10. What is the overall length of all the micro-capillaries combined?
11. What is an example of capillaries in your life?

Answers

1. A laboratory in which doctors work with and analyze blood.
2. Capillary means "strand of hair", and are very thin tubes through which liquid flows.
3. The type of liquid in the capillaries, the density of the liquid, the diameter of the capillary, and the acceleration of gravity on the given planet.
4. As the gravity on the planet increases, the height of the liquid in the capillaries decreases.
5. There are capillaries in the cloth and the water moves through the capillaries until it drips out due to the force of gravity.
6. A perpetual motion machine is a machine that can work infinitely.
7. The system doesn't work because according to the laws of physics, a perpetual motion machine can't exist.
8. Because of the hydrophilic phenomenon in a capillary.
9. There are about 10 billion micro-capillaries in the human body.
10. The overall length is between 80 and 100 thousand kilometers.
11. One example is wet cloths.

Conversation

Two celestial bodies, Earth and Moon, are neighbors. They are both very beautiful – one is blue and the other is silver. It is known that Moon revolves around Earth, looking at Earth from different sides, and Moon wants to talk with her neighbor. However, Earth is always busy, and can't give Moon her attention.

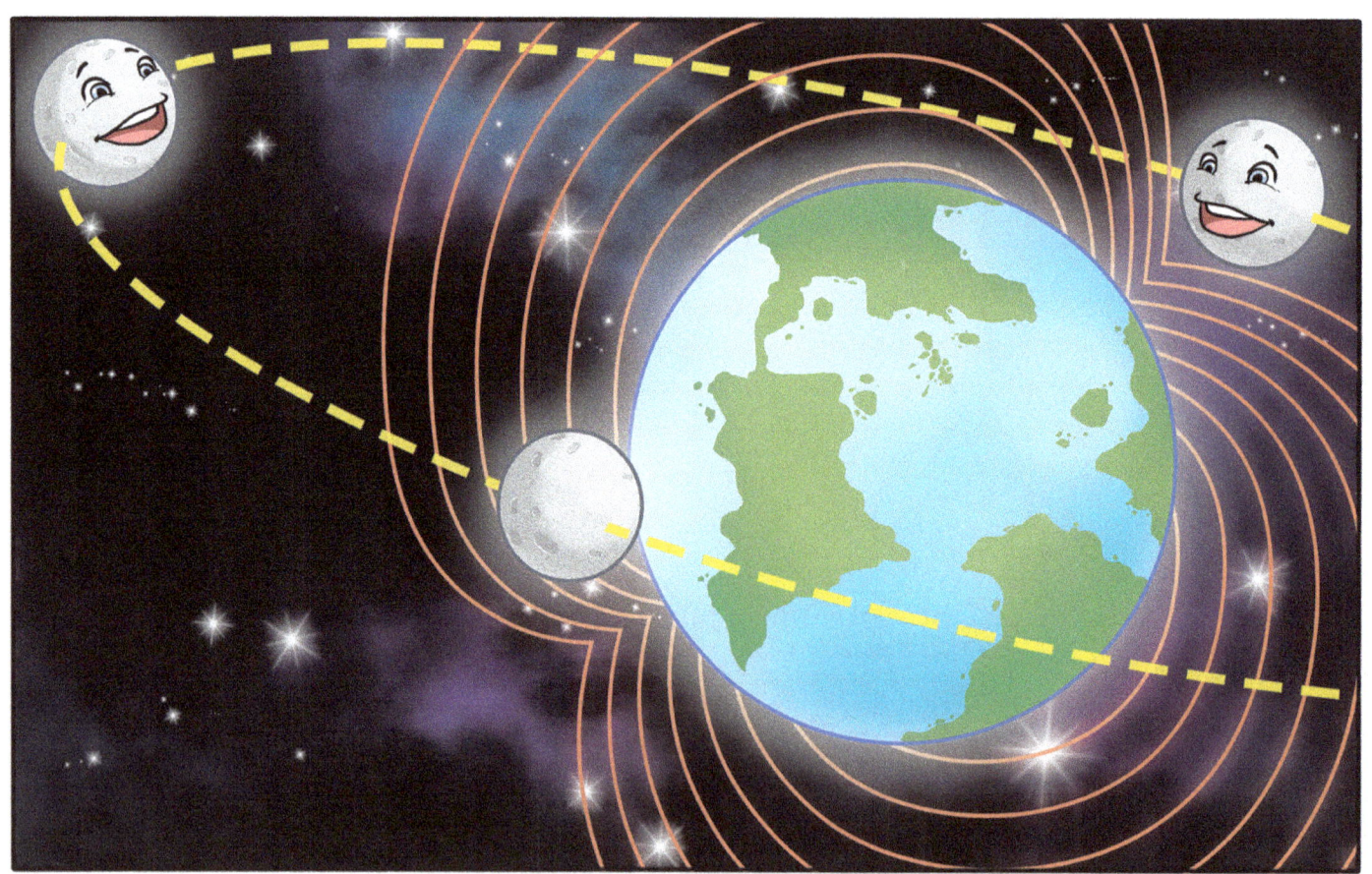

Moon: Why are you always so busy, neighbor?

Earth: There's always so much to do and worry about: saving people from disaster, feeding people, giving them clean air and comfortable atmospheric temperatures, and helping with some other unusual things that happen...

Moon: I can see that. You used to look so good, but recently you stopped dressing up and have been covered in clouds and smog.

Earth: You're right, neighbor. I have to start looking after myself, because humans don't even appreciate all my work; those egotistical creatures only think about themselves and destroy my beauty!

Moon: Very true. They built a bunch of things and used up so much metal and cement that your climate and personality are changing completely.

Earth: You know, I've realized that I'm starting to demagnetize. Scientists have been measuring my magnetic field. Guess what they saw?

Moon: What did they see?

Earth: It's shrinking.

Moon: No way!

Earth: Yes. I know what I'm talking about. I'm not sure what processes are occurring inside me, but my magnetic field is definitely decreasing.

Moon: You know, from the outside it's barely noticeable. Do you really not know what's happening inside you?

Earth: I only know that my insides are very complicated. There's the mantle, which is a layer from about 40 kilometers below sea level to about 2900 kilometers deep, where the temperature rises to 4,500 degrees Celsius. Even deeper is the outer core, which is in a melted state and has a radius of about 2200 kilometers. The flowing of the melted iron possibly creates my magnetic field. And in the center there's a core of about 1250 kilometers in radius, consisting of substances with densities of 13 grams per centimeter cubed.

Moon: Your life is much harder than it seems!

Earth: And what about you? Anything new going on in your life?

Moon: Everything used to be stable, but recently some scientists found water here. It's possible that people from Earth may start migrating to me. My life might become more interesting, as long as they do not harm me like they have you!

Earth: They are so unappreciative! I give them clean air, water, food, and even a magnetic field that protects them from many types of radiation, and they repay me by spoiling my gifts and planning to move elsewhere. Do they believe that they will live a safer life somewhere else? I guess we'll see!

Moon: It's strange that, despite the fact that I am the closest, most well-known, and most researched planetary body, scientists still know very little about me. The most detailed maps of me capture only about a quarter of my surface, and most of that area is near my equator.

Earth: On March 5th, 1998, the "Lunar Prospector", an American spacecraft designed to orbit the Moon and gather data about it, took many high-quality pictures of ice on the poles: on the North pole there was about 10-20 square kilometers of ice and on the South pole there was between 5-20 square kilometers of ice.

Moon: Huh. How long has that ice been there?

Earth: It's approximately 2 billion years old.

Moon: That means it appeared long after my birth. It probably came from comets that came from the depths of space.

Earth: These comets were commonly 90% ice, and usually would melt in my atmosphere, but since you have no atmosphere, they were able to hit your surface and did not melt.

Moon: And the fact that they hit my poles was probably just coincidence.

Earth: The poles have temperatures of negative 250 degrees Celsius, so it is not just coincidence.

Moon: This allowed the ice to be preserved until now.

Earth: In 2000, the European Space Agency planned to send a robot that would study the ice on the Moon. This mission was postponed and now, missions investigating the ice continue.

Moon: Do you know what they're planning to do in the future?

Earth: Most likely, they're planning to build a station on the Moon for astronauts and people who want to visit there.

Moon: And they will take some water from Earth, of course.

Earth: The issue is that transporting just one liter of my water costs around ten thousand dollars.

Moon: I don't know what a dollar is, but that seems expensive to me. Meanwhile, I have three hundred million tons of frozen water lying around. Just ten percent of that amount is enough for a population of two thousand people to use for a hundred years.

Earth: Furthermore, they can use the oxygen from the ice to supply the station with breathable air and hydrogen to use for fuel for their ships to make a trip back to Earth.

Moon: The first people will be living underground to protect themselves from radioactive waves and meteorites falling on the surface of the Moon.

Earth: That's horrible! They won't even be able to warm themselves in the sun!

Moon: They can warm themselves as much as they want, as long as they are in spacesuits.

Earth: You have terrible living conditions for these poor astronauts!

Moon: What's the point of having good conditions, when they will ruin it like they did on Earth?

Earth: That's true...they are so inconsiderate; they have no care for the future of their children! We need to teach them a lesson! What else can you make them learn?

Moon: In order for the lunar inhabitants to survive, they need physicists and biologists to design systems to control atmospheric pressure, temperature, the humidity of the air, food storage, and waste disposal. They also must create a limited magnetic field, similar to the one that you have on Earth.

Earth: Why can't they stay on Earth? Why do they need these adventures to the Moon?

Moon: They are bored of sitting around. They want to fly to another planet!

Earth: I don't care about them flying to other planets. I just worry about them!

Moon: To be fair, flying from here to Mars will cost them half of what it usually does from Earth.

Earth: Why is that?

Moon: Think about it: if they start from the Moon, they will have to overcome a much smaller force of gravity and on the Moon, there is no air resistance. These are very good reasons to organize spaceports on the Moon.

Earth: It's obvious that you don't love them as much as I do, because your attraction to them is much weaker than mine!

Moon: If they decide to build a huge astronomical observatory, I will not hinder them because I don't have an atmosphere, or ionosphere, or even a magnetic field like you do.

Earth: You don't understand! My atmosphere, ionosphere, and even my magnetic field are not a hindrance! They protect my people from the dangers of the cosmos!

Moon: Perhaps...but at least I will never be bored with them around.

Earth: I believe they'll return to me anyway!

Moon: Not only will they return, but also I think they will love you much more. Traveling to the Moon will be very useful.

Earth: We'll have to wait and see.

Dear parents,

For your children to understand this story clearly, we prepared investigative questions and answers for them. You could use the answers if you need them to discuss "Conversation" with your children. But first ask them to try to find the answers in the story.

Dear Children! Are you a detective? Find the answers to the questions in the text.

Questions

1. What is revolving around Earth?
2. Which celestial body, the Earth or the Moon, has a magnetic field?
3. What is inside Earth?
4. How was the magnetic field around Earth created?
5. What is the temperature inside Earth?
6. Is there water on the Moon?
7. Does the Moon have an atmosphere?
8. What is the temperature on the Moon?
9. What is a comet?
10. What is a comet composed of?
11. Could people live on the Moon? Why is it better on the Earth for them?
12. Why is it better to travel into space from the Moon than from Earth?

Answers

1. The Moon.
2. The Earth.
3. It has several layers: the inner core, the outer core, the mantle, and the surface layer.
4. Inside Earth there's molten iron that's constantly moving, which creates the magnetic field around Earth.
5. The temperature is 4,500 degrees Celsius and more as you go deeper inside the Earth.
6. Yes, although it's in the form of ice because of the extremely low temperature.
7. No, it doesn't have an atmosphere.
8. On the poles of the Moon, it's negative 250 degrees Celsius.
9. A comet is a large object that moves along its own orbit in space and sometimes collides with other objects in space.
10. Comets are large stones surrounded by ice.
11. They could live on the Moon if people are able to create conditions similar to those on Earth. Otherwise, they would have to permanently live in space suits. Earth has an atmosphere and magnetic field, which protect people from radiation and objects coming from space. Additionally, the Earth has resources such as food and water that humans need to survive.
12. The Moon has a much smaller gravitational field. Also, the Moon has no atmosphere and magnetic field.

Dear Readers! We wish you good luck in your investigation of nature's laws in the next book called

SECRETS of NATURE
— from —
D-Z

The ability to speak, to state one's thoughts clearly, to prove a point is not innate… but this ability can be very important in everyday conversation and in winning arguments.

To help you learn how to argue effectively and answer your partner's questions clearly is the aim of this book. You will find a variety of dialogues inside in which different characters try to prove their point of view on different subjects basing their arguments and statements on the laws of science.

www.ingramcontent.com/pod-product-compliance
Lightning Source LLC
Chambersburg PA
CBHW040544220526
45473CB00016B/3020